OFFICIAL SQA PAST PAPERS WITH ANSWERS

## INTERMEDIATE 2 | UNITS 1, 2 & 3

# MATHEMATICS
# 2006-2009

 SQA

 BrightRED PUBLISHING

© Scottish Qualifications Authority

First exam published in 2006.
Published by Bright Red Publishing Ltd, 6 Stafford Street, Edinburgh EH3 7AU
tel: 0131 220 5804 fax: 0131 220 6710 info@brightredpublishing.co.uk  www.brightredpublishing.co.uk

ISBN 978-1-84948-046-8

A CIP Catalogue record for this book is available from the British Library.

Bright Red Publishing is grateful to the copyright holders, as credited on the final page of the book, for permission to use their material.
Every effort has been made to trace the copyright holders and to obtain their permission for the use of copyright material.
Bright Red Publishing will be happy to receive information allowing us to rectify any error or omission in future editions.

[BLANK PAGE]

# X100/201

NATIONAL
QUALIFICATIONS
2006

FRIDAY, 19 MAY
1.00 PM – 1.45 PM

MATHEMATICS
INTERMEDIATE 2
Units 1, 2 and 3
Paper 1
(Non-calculator)

**Read carefully**

1 **You may NOT use a calculator.**

2 Full credit will be given only where the solution contains appropriate working.

3 Square-ruled paper is provided.

SCOTTISH
QUALIFICATIONS
AUTHORITY

©

**FORMULAE LIST**

The roots of $ax^2 + bx + c = 0$ are $x = \dfrac{-b \pm \sqrt{(b^2 - 4ac)}}{2a}$

Sine rule: $\dfrac{a}{\sin A} = \dfrac{b}{\sin B} = \dfrac{c}{\sin C}$

Cosine rule: $a^2 = b^2 + c^2 - 2bc \cos A$ or $\cos A = \dfrac{b^2 + c^2 - a^2}{2bc}$

Area of a triangle: $\quad$ Area $= \frac{1}{2} ab \sin C$

Volume of a sphere: $\quad$ Volume $= \frac{4}{3} \pi r^3$

Volume of a cone: $\quad$ Volume $= \frac{1}{3} \pi r^2 h$

Volume of a cylinder: $\quad$ Volume $= \pi r^2 h$

Standard deviation: $\quad s = \sqrt{\dfrac{\sum (x - \bar{x})^2}{n-1}} = \sqrt{\dfrac{\sum x^2 - (\sum x)^2 / n}{n-1}}$, where $n$ is the sample size.

*Marks*

**ALL questions should be attempted.**

1. The temperature, in degrees Celsius, at mid-day in a seaside town and the sales, in pounds, of umbrellas are shown in the scattergraph below.

   A line of best fit has been drawn.

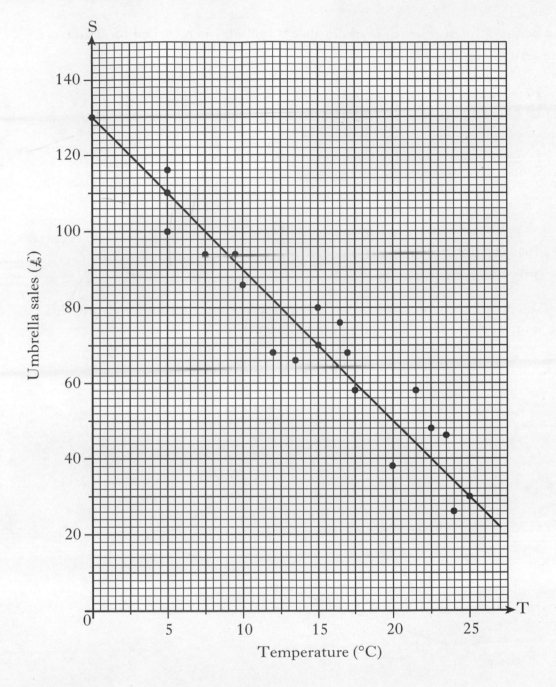

   (a) Find the equation of the line of best fit.     **3**

   (b) **Use your answer to part (a)** to predict the sales for a day when the temperature is 30 degrees Celsius.     **1**

   **[Turn over**

*Marks*

2. Multiply out the brackets and collect like terms.

$$(2y - 3)(y^2 + 4y - 1)$$

**3**

3. In a factory, the number of workers absent each day is recorded for 21 days. The results are listed below.

| | | | | | | |
|---|---|---|---|---|---|---|
| 19 | 22 | 19 | 22 | 20 | 21 | 17 |
| 19 | 21 | 16 | 20 | 19 | 18 | 18 |
| 20 | 20 | 23 | 19 | 18 | 17 | 19 |

(a) Construct a dotplot for this data.  **2**

(b) Find:
    (i)  the median;  **1**
    (ii)  the lower quartile;  **1**
    (iii)  the upper quartile.  **1**

(c) What is the probability that, on a day chosen at random from this sample, more than 18 workers were absent?  **1**

4.

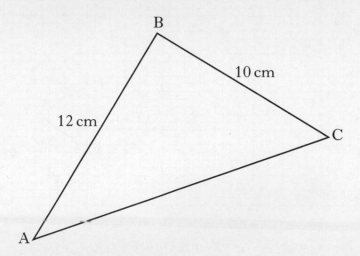

Calculate the area of triangle ABC if sin B = $\frac{2}{3}$.  **2**

*Marks*

**5.** A straight line is represented by the equation $2y + x = 6$.

    (a) Find the gradient of this line.    **2**

    (b) This line crosses the $y$-axis at $(0, c)$.
        Find the value of $c$.    **1**

**6.** Write the following in order of size, **starting with the smallest**.

$$\sin 0° \qquad \sin 30° \qquad \sin 200°$$

    Give a reason for your answer.    **2**

**7.**

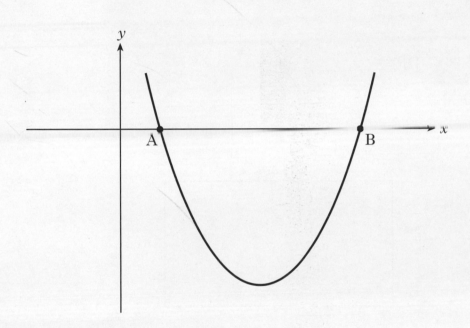

    The equation of the parabola in the above diagram is

$$y = (x - 3)^2 - 4.$$

    (a) State the coordinates of the minimum turning point of the parabola.    **2**

    (b) State the equation of the axis of symmetry of the parabola.    **1**

    (c) A is the point $(1, 0)$. State the coordinates of B.    **1**

**[Turn over for Questions 8 to 10 on** *Page*

*Marks*

**8.** The graph shown below has an equation of the form $y = \cos(x - a)^\circ$.

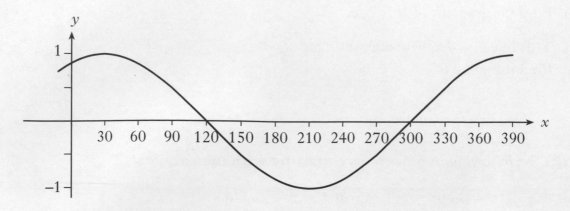

Write down the value of $a$. **1**

**9.** Evaluate

$$16^{\frac{3}{4}}.$$ **2**

**10.**

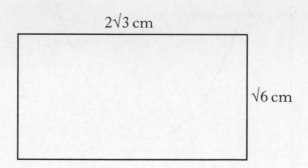

The rectangle above has length $2\sqrt{3}$ centimetres and breadth $\sqrt{6}$ centimetres.

Calculate the area of the rectangle.

Express your answer as a surd in its simplest form. **3**

*[END OF QUESTION PAPER]*

# X100/203

NATIONAL
QUALIFICATIONS
2006

FRIDAY, 19 MAY
2.05 PM – 3.35 PM

MATHEMATICS
INTERMEDIATE 2
Units 1, 2 and 3
Paper 2

**Read carefully**

1 **Calculators may be used in this paper.**

2 Full credit will be given only where the solution contains appropriate working.

3 Square-ruled paper is provided.

SCOTTISH
QUALIFICATIONS
AUTHORITY

**FORMULAE LIST**

The roots of $ax^2 + bx + c = 0$ are $x = \dfrac{-b \pm \sqrt{(b^2 - 4ac)}}{2a}$

Sine rule:    $\dfrac{a}{\sin A} = \dfrac{b}{\sin B} = \dfrac{c}{\sin C}$

Cosine rule:    $a^2 = b^2 + c^2 - 2bc \cos A$  or  $\cos A = \dfrac{b^2 + c^2 - a^2}{2bc}$

Area of a triangle:    $\text{Area} = \frac{1}{2} ab \sin C$

Volume of a sphere:    $\text{Volume} = \frac{4}{3} \pi r^3$

Volume of a cone:    $\text{Volume} = \frac{1}{3} \pi r^2 h$

Volume of a cylinder:    $\text{Volume} = \pi r^2 h$

Standard deviation:    $s = \sqrt{\dfrac{\sum (x - \bar{x})^2}{n - 1}} = \sqrt{\dfrac{\sum x^2 - (\sum x)^2 / n}{n - 1}}$, where $n$ is the sample size.

**ALL questions should be attempted.**

*Marks*

1. The value of a boat decreased from £35 000 to £32 200 in one year.

   (*a*) What was the percentage decrease? 1

   (*b*) If the value of the boat continued to fall at this rate, what would its value be after a **further** 3 years?

   Give your answer to the nearest hundred pounds. 3

2. Solve algebraically the system of equations

   $$4x + 2y = 13$$

   $$5x + 3y = 17.$$ 3

3. A child's toy is in the shape of a hemisphere with a cone on top, as shown in the diagram.

   The toy is 10 centimetres wide and 16 centimetres high.

   Calculate the volume of the toy.

   Give your answer correct to 2 significant figures.

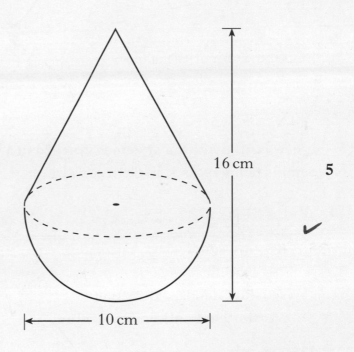

16 cm

10 cm

5

**[Turn over**

*Marks*

4. The diagram shows the base of a compact disc stand which has the shape of part of a circle.

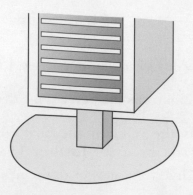

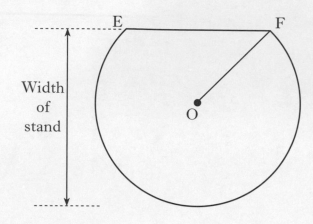

- The centre of the circle is O.
- EF is a chord of the circle.
- EF is 18 centimetres.
- The radius, OF, of the circle is 15 centimetres.

Find the width of the stand.

4

5. A new central heating system is installed in a house.
Sample temperatures, in degrees Celsius, are recorded below.

19    21    23    21    19    20

(a) For this sample data, calculate:

  (i) the mean;      **1**

  (ii) the standard deviation.      **3**

**Show clearly all your working.**

The target temperature for this house is 20 °Celsius. The system is judged to be operating effectively if the mean temperature is within 0·6 °Celsius of the target temperature **and** the standard deviation is less than 2 °Celsius.

(b) Is the system operating effectively?

**Give reasons for your answer.**      2

*Marks*

**6.** Factorise

$$4p^2 - 49.$$

2

**7.** Express

$$\frac{3}{(x+1)} - \frac{1}{(x-2)} \quad , \quad x \neq -1, \quad x \neq 2$$

as a single fraction in its simplest form.

3

**8.** The diagram shows the penalty area in a football pitch.
All measurements are given in yards.

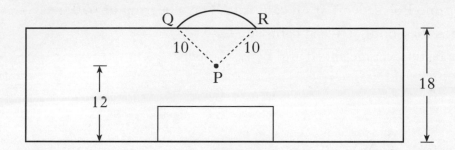

The penalty spot is marked at point P.

QR is an arc of a circle, centre P, radius 10 yards.

The width of the penalty area is 18 yards and the distance of the penalty spot from the goal line is 12 yards, as shown.

(a) Calculate the size of angle QPR.

3

(b) Calculate the length of arc QR.

2

**9.** Change the subject of the formula

$$\frac{x}{c} + a = b$$

to $x$.

2

**[Turn over**

*Marks*

10. The diagram below shows the position of three campsites A, B and C.

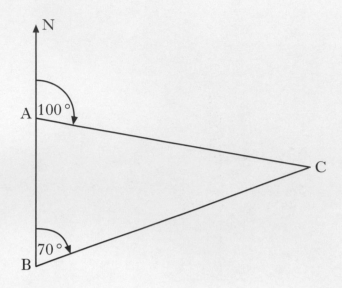

Alan sets off from campsite A on a bearing of 100° at an average speed of 5·6 kilometres per hour.

At the same time Bob sets off from campsite B on a bearing of 070°.

**After 3 hours** they both arrive at campsite C.

Who has the faster average speed and by how much?          5

11. A cuboid is shown below.

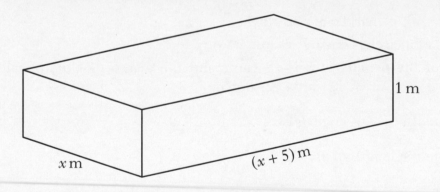

It has length $(x + 5)$ metres, breadth $x$ metres, height 1 metre and volume 24 cubic metres.

(a) Show that

$$x^2 + 5x - 24 = 0.$$          2

(b) Using the equation in part (a), find the breadth of the cuboid.          3

**12.** The arms on a wind turbine rotate at a steady rate.

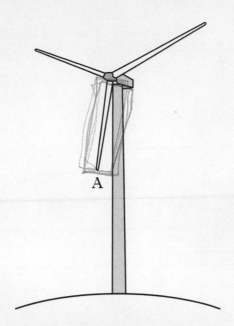

The height, $h$ metres, of a point A above the ground at time $t$ seconds is given by the equation

$$h = 8 + 4 \sin t°$$

(a) Calculate the height of point A at time 30 seconds.　　　　　　　　　　2

(b) Find the **two** times during the first turn of the arms when point A is at a height of 10·5 metres.　　　　　　　　　　4

*[END OF QUESTION PAPER]*

[BLANK PAGE]

**INTERMEDIATE 2**

# 2007

[BLANK PAGE]

# X100/201

NATIONAL
QUALIFICATIONS
2007

TUESDAY, 15 MAY
1.00 PM – 1.45 PM

MATHEMATICS
INTERMEDIATE 2
Units 1, 2 and 3
Paper 1
(Non-calculator)

**Read carefully**

1 **You may NOT use a calculator.**

2 Full credit will be given only where the solution contains appropriate working.

3 Square-ruled paper is provided.

SCOTTISH
QUALIFICATIONS
AUTHORITY

## FORMULAE LIST

The roots of $ax^2 + bx + c = 0$ are $x = \dfrac{-b \pm \sqrt{(b^2 - 4ac)}}{2a}$

Sine rule:    $\dfrac{a}{\sin A} = \dfrac{b}{\sin B} = \dfrac{c}{\sin C}$

Cosine rule:    $a^2 = b^2 + c^2 - 2bc \cos A$   or   $\cos A = \dfrac{b^2 + c^2 - a^2}{2bc}$

Area of a triangle:    $\text{Area} = \frac{1}{2}ab \sin C$

Volume of a sphere:    $\text{Volume} = \frac{4}{3}\pi r^3$

Volume of a cone:    $\text{Volume} = \frac{1}{3}\pi r^2 h$

Volume of a cylinder:    $\text{Volume} = \pi r^2 h$

Standard deviation:    $s = \sqrt{\dfrac{\sum(x - \bar{x})^2}{n-1}} = \sqrt{\dfrac{\sum x^2 - (\sum x)^2 / n}{n-1}}$, where $n$ is the sample size.

*Marks*

**ALL questions should be attempted.**

1. The table below shows the results of a survey of First Year pupils.

|       | *Wearing a blazer* | *Not wearing a blazer* |
|-------|:---:|:---:|
| *Boys*  | 40 | 22 |
| *Girls* | 29 | 9  |

What is the probability that a pupil, chosen at random from this sample, will be a girl wearing a blazer?

1

2.

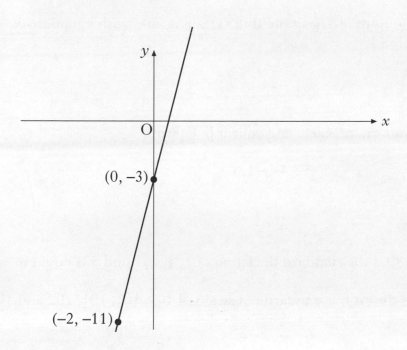

Find the equation of the straight line passing through the points (0, −3) and (−2, −11).

3

**[Turn over**

*Marks*

3.    A tin of tuna is in the shape of a cylinder.

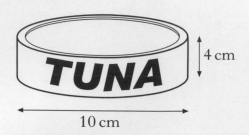

It has diameter 10 centimetres and height 4 centimetres.

Calculate its volume.

**Take π = 3·14.**

2

4.    Find the point of intersection of the straight lines with equations $x + 2y = -5$ and $3x - y = 13$.

4

5.    Multiply out the brackets and collect like terms.

$$(x + 3)(x^2 + 4x - 12)$$

3

6.    (*a*)  Show that the standard deviation of 1, 1, 1, 2 and 5 is equal to $\sqrt{3}$.

3

     (*b*)  **Write down** the standard deviation of 101, 101, 101, 102 and 105.

1

*Marks*

**7.** The graph shown below is part of the parabola with equation $y = 8x - x^2$.

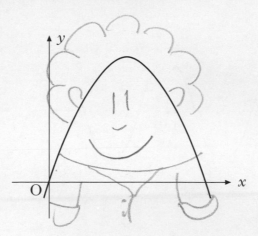

(a) By factorising $8x - x^2$, find the roots of the equation

$$8x - x^2 = 0.$$    **2**

(b) State the equation of the axis of symmetry of the parabola.    **1**

(c) Find the coordinates of the turning point.    **2**

$$8x - x^2 = 0$$

$$(8\ 4x - x)(\quad + \quad x)$$

**8.** Given that

$$\cos 60° = 0.5,$$

what is the value of $\cos 240°$?    **1**

**9.** A right-angled triangle is shown below.

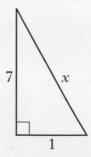

Using Pythagoras' Theorem, find $x$.

Express your answer as a surd in its simplest form.    **3**

**[Turn over for Questions 10 and 11 on *Page six***

*Marks*

10. (a) Part of the graph of $y = \cos ax°$ is shown below.

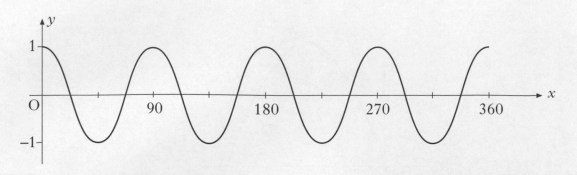

State the value of $a$.                                                                                        1

(b) Part of the graph of $y = \tan bx°$ is shown below.

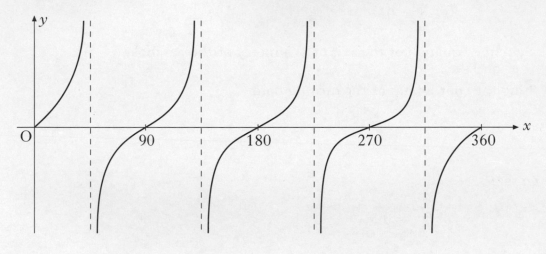

State the value of $b$.                                                                                        1

11. A straight line is represented by the equation $y = ax + b$.

Sketch a possible straight line graph to illustrate this equation when $a = 0$ and $b > 0$.                                                                       2

[*END OF QUESTION PAPER*]

# X100/203

NATIONAL
QUALIFICATIONS
2007

TUESDAY, 15 MAY
2.05 PM – 3.35 PM

MATHEMATICS
INTERMEDIATE 2
Units 1, 2 and 3
Paper 2

**Read carefully**

1 **Calculators may be used in this paper.**

2 Full credit will be given only where the solution contains appropriate working.

3 Square-ruled paper is provided.

SCOTTISH
QUALIFICATIONS
AUTHORITY

**FORMULAE LIST**

The roots of $ax^2 + bx + c = 0$ are $x = \dfrac{-b \pm \sqrt{(b^2 - 4ac)}}{2a}$

Sine rule:    $\dfrac{a}{\sin A} = \dfrac{b}{\sin B} = \dfrac{c}{\sin C}$

Cosine rule:    $a^2 = b^2 + c^2 - 2bc \cos A$ or $\cos A = \dfrac{b^2 + c^2 - a^2}{2bc}$

Area of a triangle:    Area $= \frac{1}{2}ab \sin C$

Volume of a sphere:    Volume $= \frac{4}{3}\pi r^3$

Volume of a cone:    Volume $= \frac{1}{3}\pi r^2 h$

Volume of a cylinder:    Volume $= \pi r^2 h$

Standard deviation:    $s = \sqrt{\dfrac{\sum (x - \bar{x})^2}{n-1}} = \sqrt{\dfrac{\sum x^2 - (\sum x)^2 / n}{n-1}}$, where $n$ is the sample size.

**ALL questions should be attempted.**

*Marks*

1. Ian's annual salary is £28 400. His boss tells him that his salary will increase by 2·3% per annum.

   What will Ian's annual salary be after 3 years?

   Give your answer to the nearest pound.

   3

2. The diagram below shows a sector of a circle, centre C.

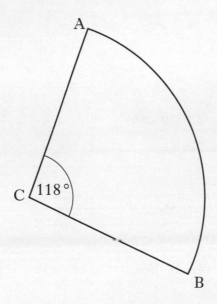

   The radius of the circle is 10·5 centimetres and angle ACB is 118°.

   Calculate the length of arc AB.

   3

   **[Turn over**

*Marks*

**3.** This back-to-back stem and leaf diagram shows the results for a class in a recent mathematics examination.

| Girls | | Boys |
|---|---|---|
| | 1 | 3 |
| 9 | 4 | 7  9 |
| 8  7  4  3  2  2 | 5 | 2  3  4  4  6  6  7  9 |
| 9  4 | 6 | 3 |
| 9  6  3 | 7 | 4  8 |
| 8  1 | 8 | 7 |

n = 15          n = 14

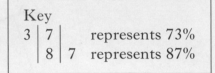

Key
3 | 7       represents 73%
   8 | 7   represents 87%

(*a*) A boxplot is drawn to represent one set of data.

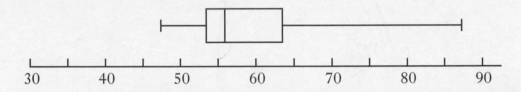

Does the boxplot above represent the girls' data or the boys' data?
**Give a reason for your answer**.                                                        1

(*b*) For the **other** set of data, find:
   (i)   the median;                                                                        1
   (ii)  the lower quartile;                                                                1
   (iii) the upper quartile.                                                                1

(*c*) Use the answers found in part (*b*) to construct a second boxplot.                    2

(*d*) Make an appropriate comment about the distribution of data in the two sets.          1

*Marks*

4.

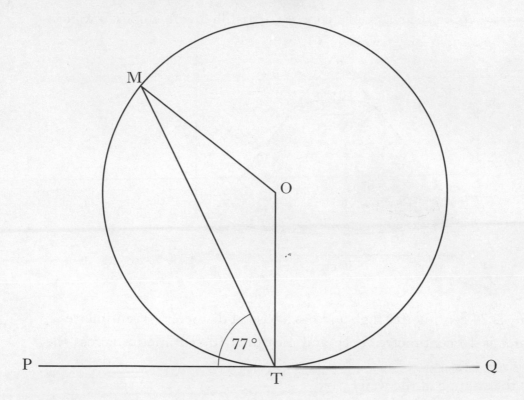

The tangent PQ touches the circle, centre O, at T.
Angle MTP is 77°.

(*a*)  Calculate the size of angle MOT.                                            **2**

(*b*)  The radius of the circle is 8 centimetres.
       Calculate the length of chord MT.                                           **3**

**[Turn over**

*Marks*

**5.** A glass ornament in the shape of a cone is partly filled with coloured water.

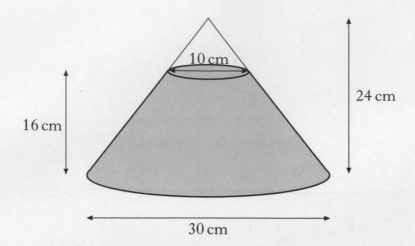

The cone is 24 centimetres high and has a base of diameter 30 centimetres.

The water is 16 centimetres deep and measures 10 centimetres across the top.

What is the volume of the water?

Give your answer correct to 2 significant figures.     **5**

**6.** Tasnim rolls a standard dice with faces numbered 1 to 6.

The probability that she gets a number less than 7 is

$$
\begin{array}{ll}
A & 0 \\
B & \frac{1}{7} \\
C & \frac{1}{6} \\
D & 1.
\end{array}
$$

Write down the letter that corresponds to the correct probability.     **1**

**7.** (*a*) Factorise **fully**

$$2x^2 - 18.$$     **2**

(*b*) Simplify

$$\frac{(2x+5)^2}{(2x-1)(2x+5)}.$$     **1**

*Marks*

**8.** Solve the equation

$$2x^2 - 6x - 5 = 0,$$

giving the roots correct to one decimal place.    **4**

**9.** The diagram shows two blocks of flats of equal height.

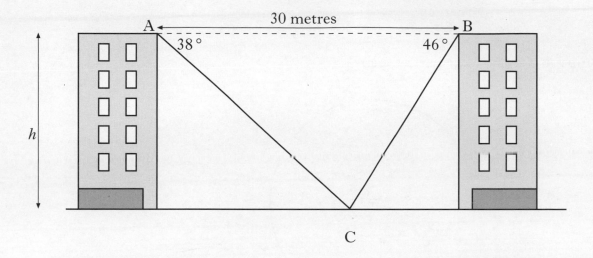

A and B represent points on the top of the flats and C represents a point on the ground between them.

To calculate the height, $h$, of each block of flats, a surveyor measures the angles of depression from A and B to C.

From A, the angle of depression is $38°$.
From B, the angle of depression is $46°$.
The distance AB is 30 metres.

Calculate the height, $h$, in metres.    **5**

**10.** Express $\dfrac{5p^2}{8} \div \dfrac{p}{2}$ as a fraction in its simplest form.    **3**

**11.** Change the subject of the formula

$$K = \frac{m^2 n}{p}$$

to $m$.    **3**

**[Turn over for Questions 12, 13 and 14 on *Page eight***

*Marks*

**12.** Simplify the expression below, giving your answer with a positive power.

$$m^5 \times m^{-8}$$

2

**13.** Solve the equation

$$5 \tan x° - 6 = 2, \qquad 0 \leq x < 360.$$

3

**14.** A mirror is shaped like part of a circle.

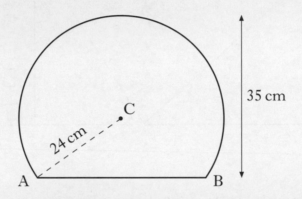

The radius of the circle, centre C, is 24 centimetres.
The height of the mirror is 35 centimetres.

Calculate the length of the base of the mirror, represented in the diagram by AB.

3

[*END OF QUESTION PAPER*]

# 2008

[BLANK PAGE]

# X100/201

NATIONAL
QUALIFICATIONS
2008

TUESDAY, 20 MAY
1.00 PM – 1.45 PM

MATHEMATICS
INTERMEDIATE 2
Units 1, 2 and 3
Paper 1
(Non-calculator)

**Read carefully**

1 You may **NOT** use a calculator.

2 Full credit will be given only where the solution contains appropriate working.

3 Square-ruled paper is provided.

**FORMULAE LIST**

The roots of $ax^2 + bx + c = 0$ are $x = \dfrac{-b \pm \sqrt{\left(b^2 - 4ac\right)}}{2a}$

Sine rule: $\dfrac{a}{\sin A} = \dfrac{b}{\sin B} = \dfrac{c}{\sin C}$

Cosine rule: $a^2 = b^2 + c^2 - 2bc \cos A$  or  $\cos A = \dfrac{b^2 + c^2 - a^2}{2bc}$

Area of a triangle:     Area $= \frac{1}{2}ab \sin C$

Volume of a sphere:     Volume $= \frac{4}{3}\pi r^3$

Volume of a cone:     Volume $= \frac{1}{3}\pi r^2 h$

Volume of a cylinder:     Volume $= \pi r^2 h$

Standard deviation:     $s = \sqrt{\dfrac{\sum (x - \bar{x})^2}{n-1}} = \sqrt{\dfrac{\sum x^2 - (\sum x)^2 / n}{n-1}}$, where $n$ is the sample size.

*Marks*

**ALL questions should be attempted.**

1. A straight line has equation $y = 4x + 5$.

   State the gradient of this line.　　　　　　　　　　　　　　　　　　　　1

2. Multiply out the brackets and collect like terms.

   $$(3x + 2)(x - 5) + 8x$$　　　　　　　　　　　　　　　　　　　　　　3

3. The stem and leaf diagram shows the number of points gained by the football teams in the Premiership League in a season.

   ```
   3 | 3 3 3 9
   4 | 1 4 5 5 7 8
   5 | 0 2 3 3 6 6
   6 | 0
   7 | 5 9
   8 |
   9 | 0
   ```

   　　n = 20　　　　　　　　4 | 1 represents 41 points

   (a) Arsenal finished 1st in the Premiership with 90 points.

   　　In what position did Southampton finish if they gained 47 points?　　1

   (b) What is the probability that a team chosen at random scored less than 44 points?　　1

4. (a) Factorise

   $$x^2 - y^2.$$　　　　　　　　　　　　　　　　　　　　　　　　　1

   (b) Hence, or otherwise, find the value of

   $$9 \cdot 3^2 - 0 \cdot 7^2.$$　　　　　　　　　　　　　　　　　　　2

**[Turn over**

*Marks*

**5.** In a survey, the number of books carried by each girl in a group of students was recorded.

The results are shown in the frequency table below.

| Number of books | Frequency |
|-----------------|-----------|
| 0 | 1 |
| 1 | 2 |
| 2 | 3 |
| 3 | 5 |
| 4 | 5 |
| 5 | 6 |
| 6 | 2 |
| 7 | 1 |

   (*a*) Copy this frequency table and add a cumulative frequency column.     **1**

   (*b*) For this data, find:

      (i) the median;     **1**

      (ii) the lower quartile;     **1**

      (iii) the upper quartile.     **1**

   (*c*) Calculate the semi-interquartile range.     **1**

   (*d*) In the same survey, the number of books carried by each boy was also recorded.

The semi-interquartile range was 0·75.

Make an appropriate comment comparing the distribution of data for the girls and the boys.     **1**

**6.** Triangle PQR is shown below.

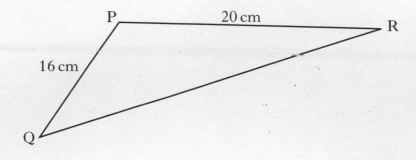

If $\sin P = \frac{1}{4}$, calculate the area of triangle PQR.     **2**

*Marks*

**7.**

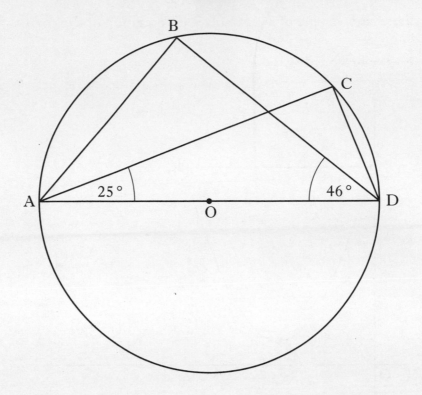

AD is a diameter of a circle, centre O.

B and C are points on the circumference of the circle.

Angle CAD = 25°.

Angle BDA = 46°.

Calculate the size of angle BAC.

3

**8.** Part of the graph of $y = a \sin bx°$ is shown in the diagram.

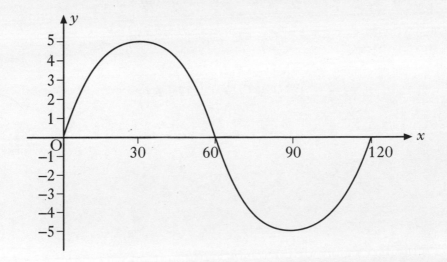

State the values of *a* and *b*.

2

**[Turn over for Questions 9 and 10 on *Page six***

*Marks*

9. The graph below shows part of a parabola with equation of the form

   $$y = (x \overset{-}{+} a)^2 + b.$$

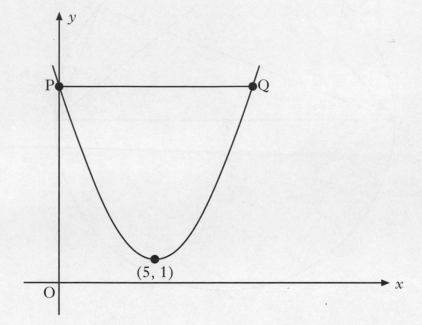

(a) State the values of $a$ and $b$.    2

(b) State the equation of the axis of symmetry of the parabola.    1

(c) The line PQ is parallel to the $x$-axis.
    Find the coordinates of points P and Q.    3

10. If $\sin x° = \dfrac{4}{5}$ and $\cos x° = \dfrac{3}{5}$, calculate the value of $\tan x°$.    2

*[END OF QUESTION PAPER]*

# X100/203

NATIONAL
QUALIFICATIONS
2008

TUESDAY, 20 MAY
2.05 PM – 3.35 PM

MATHEMATICS
INTERMEDIATE 2
Units 1, 2 and 3
Paper 2

**Read carefully**

1  **Calculators may be used in this paper.**

2  Full credit will be given only where the solution contains appropriate working.

3  Square-ruled paper is provided.

**FORMULAE LIST**

The roots of $ax^2 + bx + c = 0$ are $x = \dfrac{-b \pm \sqrt{(b^2 - 4ac)}}{2a}$

Sine rule: $\dfrac{a}{\sin A} = \dfrac{b}{\sin B} = \dfrac{c}{\sin C}$

Cosine rule: $a^2 = b^2 + c^2 - 2bc \cos A$ or $\cos A = \dfrac{b^2 + c^2 - a^2}{2bc}$

Area of a triangle: $\text{Area} = \frac{1}{2}ab \sin C$

Volume of a sphere: $\text{Volume} = \frac{4}{3}\pi r^3$

Volume of a cone: $\text{Volume} = \frac{1}{3}\pi r^2 h$

Volume of a cylinder: $\text{Volume} = \pi r^2 h$

Standard deviation: $s = \sqrt{\dfrac{\sum(x - \bar{x})^2}{n - 1}} = \sqrt{\dfrac{\sum x^2 - (\sum x)^2 / n}{n - 1}}$, where $n$ is the sample size.

**ALL questions should be attempted.**

*Marks*

1. Calculate the **compound interest** earned when £50 000 is invested for 4 years at 4·5% per annum.

   Give your answer to the nearest penny.

   **4**

2. Jim Reid keeps his washing in a basket. The basket is in the shape of a prism.

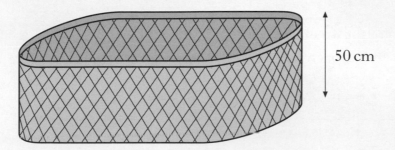

   The height of the basket is 50 centimetres.

   The cross section of the basket consists of a rectangle and two semi-circles with measurements as shown.

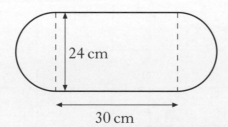

   (a) Find the volume of the basket in cubic centimetres.

   Give your answer correct to three significant figures.

   **4**

   Jim keeps his ironing in a storage box which has a volume **half** that of the basket.

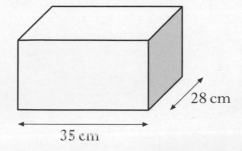

   The storage box is in the shape of a cuboid, 35 centimetres long and 28 centimetres broad.

   (b) Find the height of the storage box.

   **3**

*Marks*

**3.** The results for a group of students who sat tests in mathematics and physics are shown below.

| Mathematics (%) | 10 | 18 | 26 | 32 | 49 |
|---|---|---|---|---|---|
| Physics (%) | 25 | 35 | 30 | 40 | 41 |

(a) Calculate the standard deviation for the mathematics test.

**4**

(b) The standard deviation for physics was 6·8.

Make an appropriate comment on the distribution of marks in the two tests.

**1**

These marks are shown on the scattergraph below.

A line of best fit has been drawn.

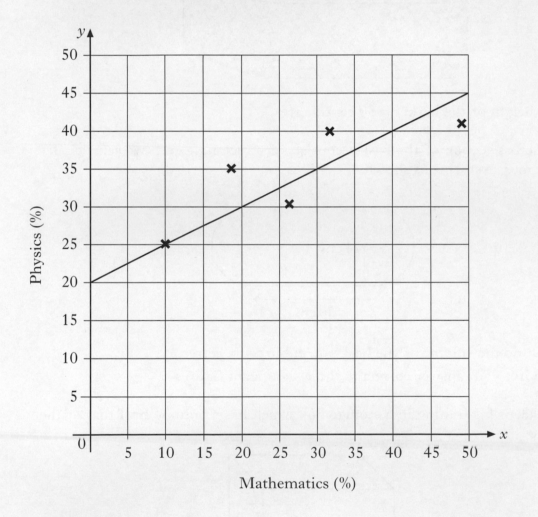

Mathematics (%)

(c) Find the equation of the line of best fit.

**3**

(d) Another pupil scored 76% in the mathematics test but was absent from the physics test.

**Use your answer to part (c)** to predict his physics mark.

**1**

*Marks*

4.  Suzie has a new mobile phone.  She is charged $x$ pence per minute for calls and $y$ pence for each text she sends.   During the first month her calls last a total of 280 minutes and she sends 70 texts.  Her bill is £52·50.

(a)  Write down an equation in $x$ and $y$ which satisfies the above condition.    **1**

The next month she reduces her bill.  She restricts her calls to 210 minutes and sends 40 texts.  Her bill is £38·00.

(b)  Write down a second equation in $x$ and $y$ which satisfies this condition.    **1**

(c)  Calculate the price per minute for a call and the price for each text sent.    **4**

5.  Triangle DEF is shown below.

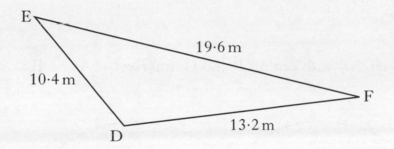

It has sides of length 10·4 metres, 13·2 metres and 19·6 metres.
Calculate the size of angle EDF.
**Do not use a scale drawing.**    **3**

6.  Solve the equation

$$5x^2 + 4x - 2 = 0,$$

giving the roots correct to 2 decimal places.    **4**

**[Turn over**

*Marks*

7. (a) Simplify

$$\frac{m^5}{m^3}.$$

1

(b) Express

$$2\sqrt{5} + \sqrt{20} - \sqrt{45}$$

as a surd in its simplest form.

3

8. Solve the equation

$$4\cos x^\circ + 3 = 0, \qquad 0 \le x \le 360.$$

3

9. Two identical circles, with centres P and Q, intersect at A and B as shown in the diagram.

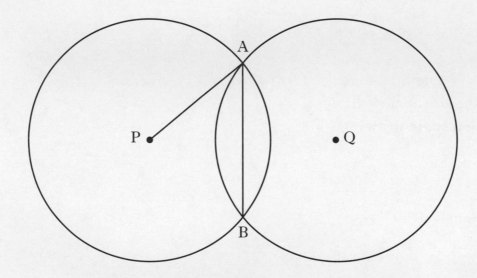

The radius of each circle is 10 centimetres.
The length of the common chord, AB, is 12 centimetres.

Calculate PQ, the distance between the centres of the two circles.

5

*Marks*

**10.** Change the subject of the formula

$$p = q + \sqrt{a}$$

to $a$.

2

**11.** Express

$$\frac{2}{a} - \frac{3}{(a+4)}, \qquad a \neq 0,\ a \neq -4,$$

as a single fraction in its simplest form.

3

*[END OF QUESTION PAPER]*

[BLANK PAGE]

INTERMEDIATE 2

# 2009

[BLANK PAGE]

# X100/201

NATIONAL
QUALIFICATIONS
2009

THURSDAY, 21 MAY
1.00 PM – 1.45 PM

## MATHEMATICS
INTERMEDIATE 2
Units 1, 2 and 3
Paper 1
(Non-calculator)

**Read carefully**

1   **You may <u>NOT</u> use a calculator.**

2   Full credit will be given only where the solution contains appropriate working.

3   Square-ruled paper is provided.

**FORMULAE LIST**

The roots of $ax^2 + bx + c = 0$ are $x = \dfrac{-b \pm \sqrt{(b^2 - 4ac)}}{2a}$

Sine rule:   $\dfrac{a}{\sin A} = \dfrac{b}{\sin B} = \dfrac{c}{\sin C}$

Cosine rule:   $a^2 = b^2 + c^2 - 2bc \cos A$   or   $\cos A = \dfrac{b^2 + c^2 - a^2}{2bc}$

Area of a triangle:     Area $= \frac{1}{2}ab \sin C$

Volume of a sphere:     Volume $= \frac{4}{3}\pi r^3$

Volume of a cone:     Volume $= \frac{1}{3}\pi r^2 h$

Volume of a cylinder:   Volume $= \pi r^2 h$

Standard deviation:     $s = \sqrt{\dfrac{\sum(x - \bar{x})^2}{n-1}} = \sqrt{\dfrac{\sum x^2 - (\sum x)^2 / n}{n-1}}$, where $n$ is the sample size.

*Marks*

**ALL questions should be attempted.**

1. The number of goals scored one weekend by each team in the Football League is shown below.

   | 0 | 1 | 1 | 2 | 1 | 0 | 0 | 5 | 0 | 1 | 3 |
   |---|---|---|---|---|---|---|---|---|---|---|
   | 0 | 2 | 2 | 1 | 1 | 3 | 0 | 0 | 2 | 4 | 1 |

   (a) Construct a dotplot for the data.

   **2**

   (b) The shape of the distribution is

   　　　　A    skewed to the right
   　　　　B    symmetric
   　　　　C    skewed to the left
   　　　　D    uniform.

   Write down the letter that corresponds to the correct shape.

   **1**

2.

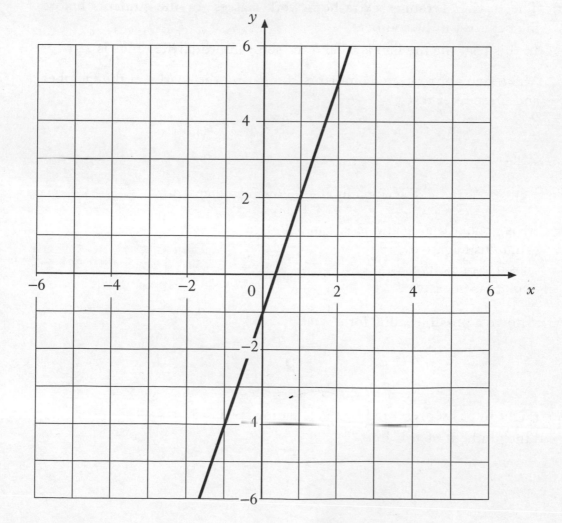

Find the equation of the straight line shown in the diagram.

**3**

*Marks*

**3.** Factorise

$$x^2 - 5x - 24.$$

2

**4.** Multiply out the brackets and collect like terms.

$$(x + 5)(2x^2 - 3x - 1)$$

3

**5.** (*a*) The marks of a group of students in their October test are listed below.

41   56   68   59   43   37   70   58   61   47   75   66

Calculate:

(i)   the median;

1

(ii)   the semi-interquartile range.

3

(*b*) The teacher arranges extra homework classes for the students before the next test in December.

In this test, the median is 67 and the semi-interquartile range is 7.

Make **two** appropriate comments comparing the marks in the October and December tests.

2

**6.** An angle, $a°$, can be described by the following statements.

- $a$ is greater than 0 and less than 360
- $\sin a°$ is negative
- $\cos a°$ is positive
- $\tan a°$ is negative

Write down a possible value for $a$.

1

**7.** A straight line is represented by the equation $x + y = 5$.
Find the gradient of this line.

2

*Marks*

**8.** Sketch the graph of $y = 4\cos 2x°$,   $0 \leq x \leq 360$.     **3**

**9.** The diagram below shows part of a parabola with equation of the form

$$y = (x + a)^2 + b.$$

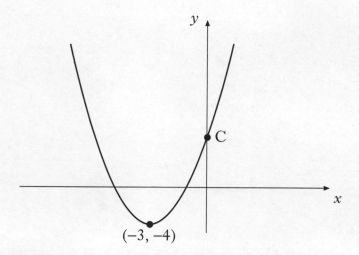

(−3, −4)

(a) Write down the equation of the axis of symmetry of the graph.     **1**

(b) Write down the equation of the parabola.     **2**

(c) Find the coordinates of C.     **2**

**10.** Simplify

$$\frac{\cos^3 x°}{1 - \sin^2 x°}.$$
     **2**

*[END OF QUESTION PAPER]*

[BLANK PAGE]

# X100/203

NATIONAL
QUALIFICATIONS
2009

THURSDAY, 21 MAY
2.05 PM – 3.35 PM

MATHEMATICS
INTERMEDIATE 2
Units 1, 2 and 3
Paper 2

**Read carefully**

1  **Calculators may be used in this paper.**

2  Full credit will be given only where the solution contains appropriate working.

3  Square-ruled paper is provided.

LI  X100/203  6/24220

**FORMULAE LIST**

The roots of $ax^2 + bx + c = 0$ are $x = \dfrac{-b \pm \sqrt{(b^2 - 4ac)}}{2a}$

Sine rule:    $\dfrac{a}{\sin A} = \dfrac{b}{\sin B} = \dfrac{c}{\sin C}$

Cosine rule:    $a^2 = b^2 + c^2 - 2bc \cos A$  or  $\cos A = \dfrac{b^2 + c^2 - a^2}{2bc}$

Area of a triangle:    Area $= \frac{1}{2}ab \sin C$

Volume of a sphere:    Volume $= \frac{4}{3}\pi r^3$

Volume of a cone:    Volume $= \frac{1}{3}\pi r^2 h$

Volume of a cylinder:    Volume $= \pi r^2 h$

Standard deviation:    $s = \sqrt{\dfrac{\sum(x - \bar{x})^2}{n-1}} = \sqrt{\dfrac{\sum x^2 - (\sum x)^2 / n}{n-1}}$, where $n$ is the sample size.

**ALL questions should be attempted.**

1. A new book "Intermediate 2 Maths is Fun" was published in 2006.
   There were 3000 sales of the book during that year.
   Sales rose by 11% in 2007 then fell by 10% in 2008.

   Were the sales in 2008 more or less than the sales in 2006?

   **You must give a reason for your answer.**  **3**

2. The heights, in centimetres, of seven netball players are given below.

   173    176    168    166    170    180    171

   For this sample, calculate:

   (*a*) the mean;  **1**

   (*b*) the standard deviation.  **3**

   **Show clearly all your working.**

   **[Turn over**

*Marks*

**3.** A company manufactures aluminium tubes.

The cross-section of one of the tubes is shown in the diagram below.

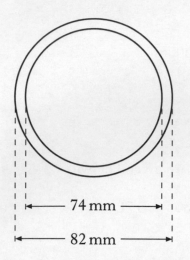
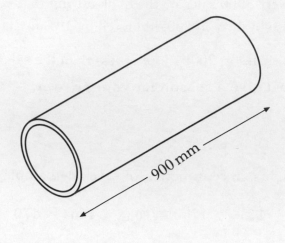

74 mm

82 mm

900 mm

The inner diameter is 74 millimetres.

The outer diameter is 82 millimetres.

The tube is 900 millimetres long.

Calculate the volume of aluminium used to make the tube.

**Give your answer correct to three significant figures.**    **5**

**4.** There are 14 cars and 60 passengers on the morning crossing of the ferry from Wemyss Bay to Rothesay. The total takings are £344·30.

(a) Let *x* pounds be the cost for a car and *y* pounds be the cost for a passenger.

Write down an equation in *x* and *y* which satisfies the above condition.    **1**

(b) There are 21 cars and 40 passengers on the evening crossing of the ferry. The total takings are £368·95.

Write down a second equation in *x* and *y* which satisfies this condition.    **1**

(c) Find the cost for a car and the cost for a passenger on the ferry.    **4**

*Marks*

5.  A pet shop manufactures protective dog collars.
    In the diagram below the shaded area represents
    one of these collars.

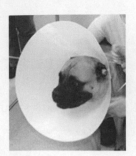

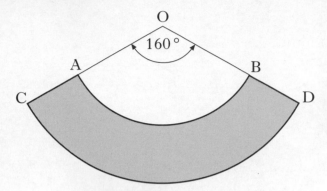

AB and CD are arcs of the circles with centres at O.

The radius, OA, is 10 inches and the radius, OC, is 18 inches.

Angle AOB is 160°.

Calculate the area of a collar.                                          **4**

6.  The Bermuda triangle is an area in the Atlantic
    Ocean where many planes and ships have
    mysteriously disappeared.

    Its vertices are at Bermuda (B), Miami (M) and
    Puerto Rico (P).

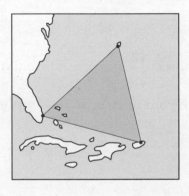

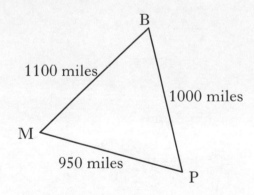

Calculate the size of angle BPM.                                        **3**

7.  Solve the equation

$$x^2 + 5x + 3 = 0,$$

giving the roots correct to one decimal place.                          **4**

**[Turn over**

*Marks*

8. Express

$$\frac{2}{x-1} + \frac{4}{x+2} \qquad x \neq 1, \ x \neq -2$$

as a single fraction in its simplest form.    **3**

9. Change the subject of the formula

$$A = \frac{1}{2}h(a+b)$$

to $h$.    **2**

10. Solve the equation

$$7\sin x° + 1 = -5, \qquad 0 \leq x \leq 360.$$    **3**

11. Express $\dfrac{12}{\sqrt{2}}$ with a rational denominator.

Give your answer in its simplest form.    **2**

12. Simplify $\quad \dfrac{ab^6}{a^3b^2}$.    **2**

*Marks*

13. For reasons of safety, a building is supported by two wooden struts, represented by DB and DC in the diagram below.

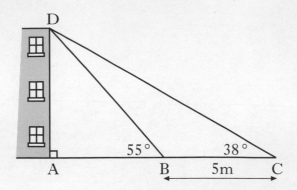

Angle ABD = 55°.

Angle BCD = 38°.

BC is 5 metres.

Calculate the height of the building represented by AD.

5

14. A railway goes through an underground tunnel.

The diagram below shows the cross-section of the tunnel. It consists of part of a circle with a horizontal base.

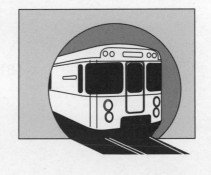

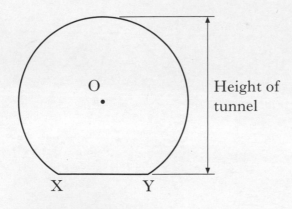

Height of tunnel

- The centre of the circle is O.
- XY is a chord of the circle.
- XY is 1·8 metres.
- The radius of the circle is 1·7 metres.

Find the height of the tunnel.

4

[*END OF QUESTION PAPER*]

[BLANK PAGE]

[BLANK PAGE]

[BLANK PAGE]